Fractions

$0, 1, 2, 3, $
numbers. <u>Whol</u>_____
units (whole things).

$\frac{1}{2}, \frac{3}{5}, \frac{5}{8}, \frac{10}{4}, \frac{6}{2}, \frac{4}{7}$ ... are fractions.
<u>Fractions</u> name parts of units.

Circle all the fractions.

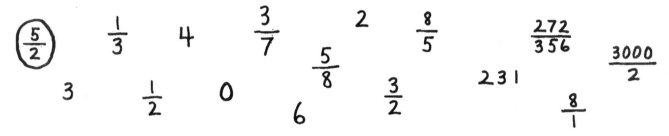

When a unit is divided into two equal parts, the parts are <u>halves</u>.
When a unit is divided into three equal parts, the parts are <u>thirds</u>.
Four equal parts are <u>fourths</u> or <u>quarters</u>.   Five equal parts are <u>fifths</u>.

Which shows halves?	Which shows thirds?	Which shows quarters?

Match.

fifths   sixths   sevenths   eighths

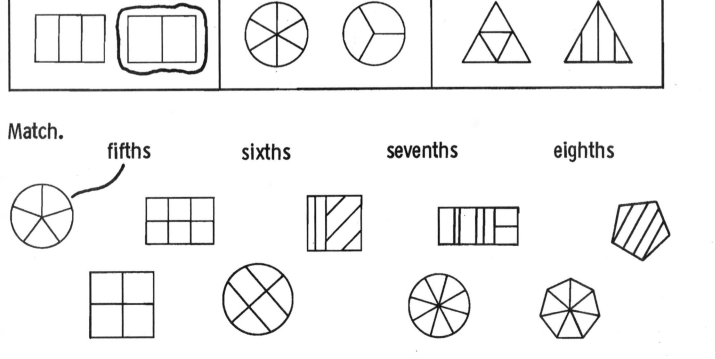

shows eight equal parts or **_eighths_** .

shows three equal parts or _____ .

shows five equal parts or _____ .

shows two equal parts or _____ .

shows four equal parts or _____ .

Fractions can be shown by dividing a unit into equal parts. Equal parts must all be the same size.

is divided into three parts. It does (does not) show thirds.

is divided into four parts. It does / does not show fourths.

is divided into five parts. It does / does not show fifths.

is divided into six parts. It does / does not show sixths.

is divided into seven parts. It does / does not show sevenths.

3

Divide into two equal parts.

This shows _2_ halves.

Divide into three equal parts.

This shows ___ thirds.

Divide into six equal parts.

This shows ___ sixths.

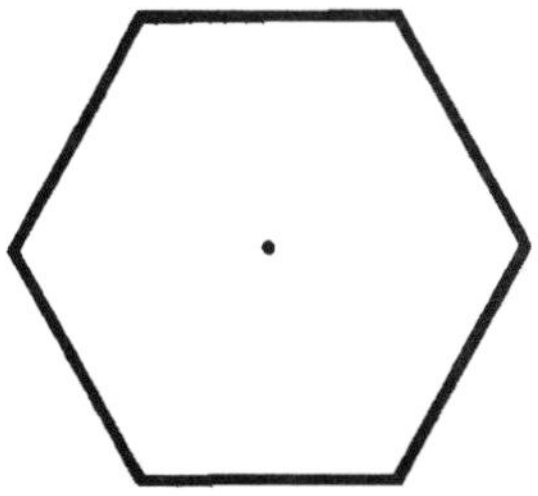

Show 6 sixths.

Show 2 halves.

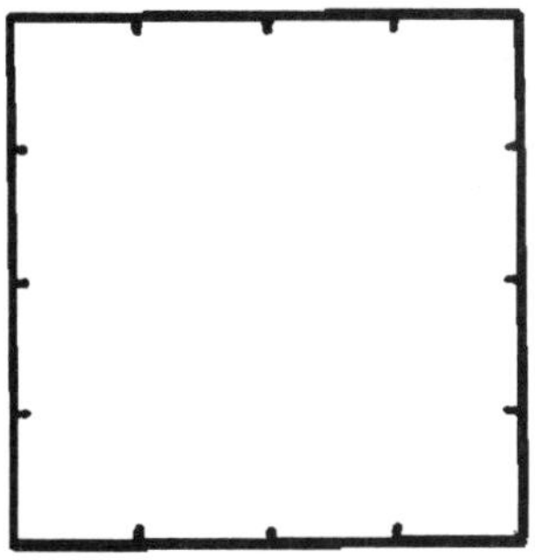

Show 4 fourths.

Show 8 eighths.

Show 16 sixteenths.

These fractions are fourths: $\frac{1}{4}$, $\frac{2}{4}$, $\frac{3}{4}$, $\frac{4}{4}$, $\frac{5}{4}$, $\frac{6}{4}$, $\frac{7}{4}$ ...

These fractions are fifths: $\frac{1}{5}$, $\frac{2}{5}$, $\frac{3}{5}$, $\frac{4}{5}$, $\frac{5}{5}$, $\frac{6}{5}$, $\frac{7}{5}$ ...

These fractions are tenths: $\frac{1}{10}$, $\frac{2}{10}$, $\frac{3}{10}$, $\frac{4}{10}$, $\frac{5}{10}$, $\frac{6}{10}$, $\frac{7}{10}$ ...

Which fractions are thirds? $\frac{1}{2}$, $\boxed{\frac{2}{3}}$, $\frac{3}{4}$, $\boxed{\frac{1}{3}}$, $\frac{1}{5}$, $\frac{7}{10}$, $\frac{3}{6}$ ...

Which fractions are sixths? $\frac{1}{3}$, $\frac{6}{6}$, $\frac{3}{6}$, $\frac{1}{4}$, $\frac{3}{4}$, $\frac{10}{6}$, $\frac{1}{5}$ ...

Which fractions are ninths? $\frac{1}{9}$, $\frac{3}{9}$, $\frac{9}{3}$, $\frac{1}{2}$, $\frac{5}{90}$, $\frac{100}{9}$, $\frac{0}{9}$ ...

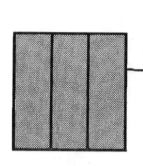

is divided into __thirds__ . It shows __3__ __thirds__ .

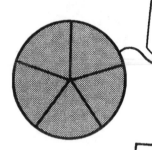

is divided into _____ . It shows ___ _____ .

is divided into _____ . It shows ___ _____ .

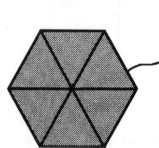

is divided into _____ . It shows ___ _____ .

Naming Fractional Parts

Name the shaded part of each figure.

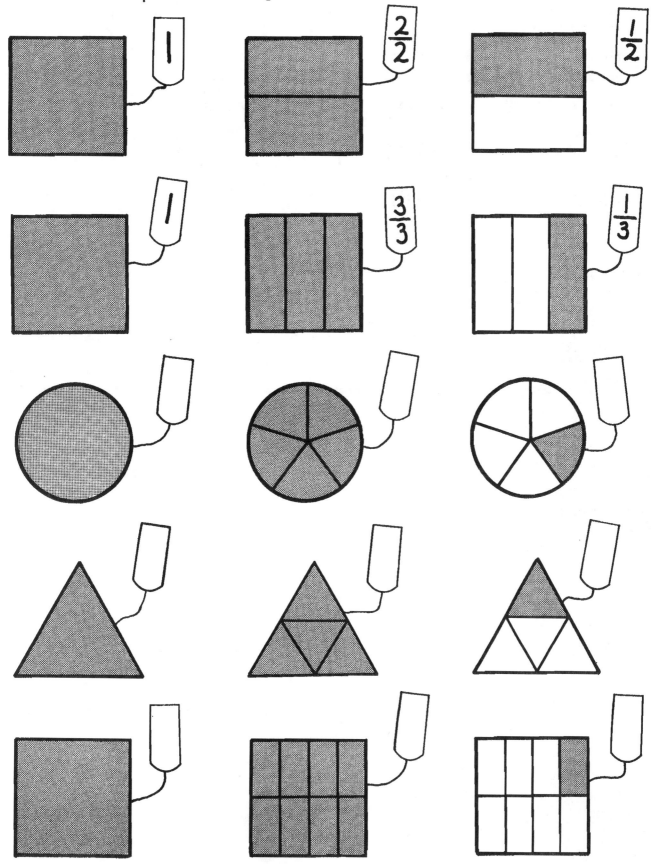

6

Name the shaded part of each figure.

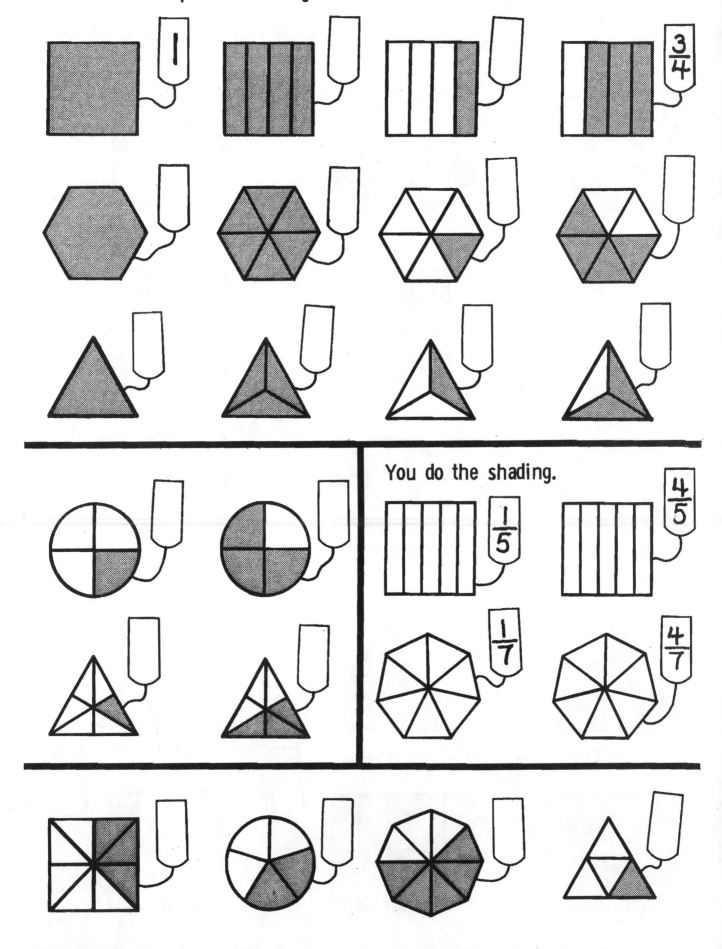

You do the shading.

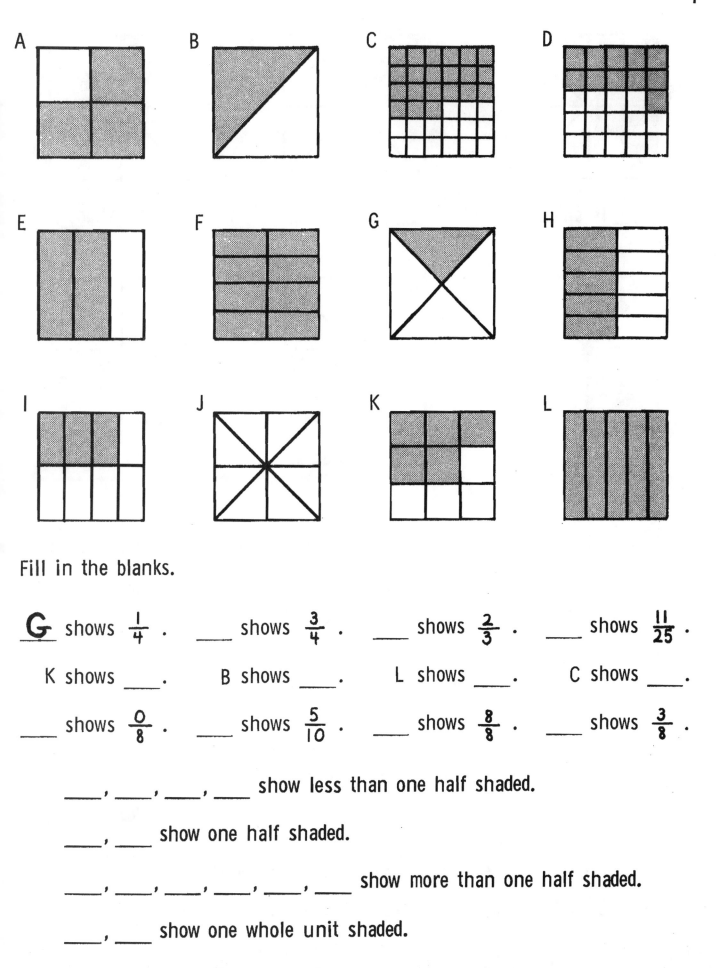

Fill in the blanks.

__G__ shows $\frac{1}{4}$. ___ shows $\frac{3}{4}$. ___ shows $\frac{2}{3}$. ___ shows $\frac{11}{25}$.

K shows ___ . B shows ___ . L shows ___ . C shows ___ .

___ shows $\frac{0}{8}$. ___ shows $\frac{5}{10}$. ___ shows $\frac{8}{8}$. ___ shows $\frac{3}{8}$.

___ , ___ , ___ , ___ show less than one half shaded.

___ , ___ show one half shaded.

___ , ___ , ___ , ___ , ___ , ___ show more than one half shaded.

___ , ___ show one whole unit shaded.

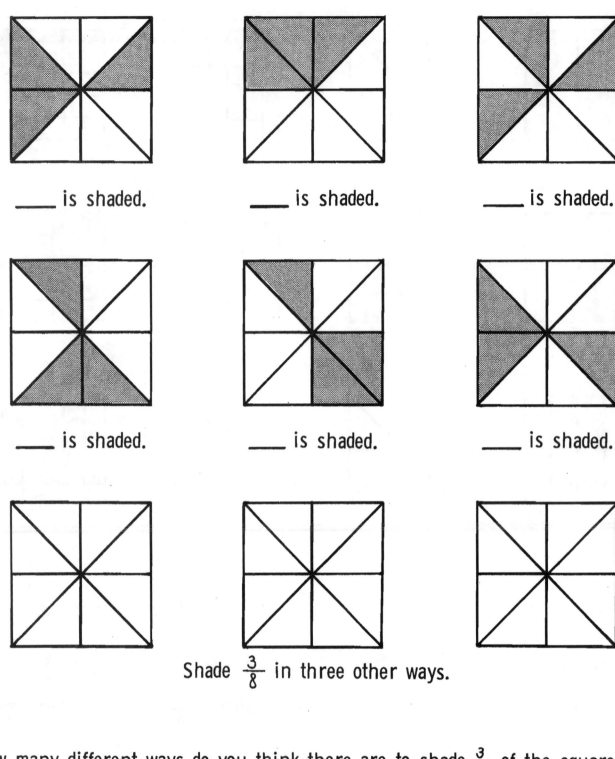

_____ is shaded.   _____ is shaded.   _____ is shaded.

_____ is shaded.   _____ is shaded.   _____ is shaded.

Shade $\frac{3}{8}$ in three other ways.

How many different ways do you think there are to shade $\frac{3}{8}$ of the square?

6   15   32   56   100

The answer is on the next page. See if you were right.

☐ I was right.   ☐ I was wrong.

Answer from last page: There are 56 different ways to shade $\frac{3}{8}$ of the square.

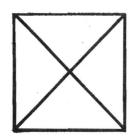

Shade $\frac{1}{4}$ of the square.

Shade $\frac{2}{5}$ of the circle.

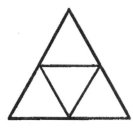

Shade $\frac{3}{4}$ of the triangle.

Shade $\frac{5}{6}$ of the hexagon.

Divide into three equal parts.
Shade $\frac{2}{3}$ of the rectangle.

Divide into fourths.
Shade $\frac{1}{4}$ of the rectangle.

Divide into four equal parts.
Shade $\frac{4}{4}$ of the triangle.

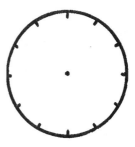

Divide into sixths.
Shade $\frac{3}{6}$ of the circle.

10

What fraction is shaded? What fraction is not shaded?

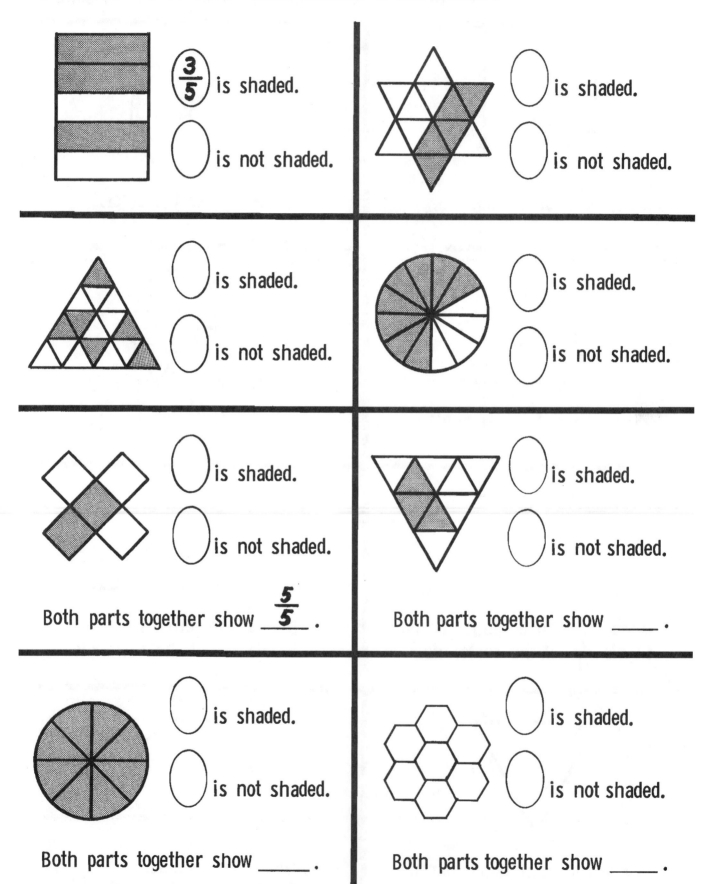

$\frac{3}{5}$ is shaded.

◯ is not shaded.

◯ is shaded.

◯ is not shaded.

◯ is shaded.

◯ is not shaded.

◯ is shaded.

◯ is not shaded.

◯ is shaded.

◯ is not shaded.

Both parts together show $\frac{5}{5}$.

◯ is shaded.

◯ is not shaded.

Both parts together show _____ .

◯ is shaded.

◯ is not shaded.

Both parts together show _____ .

◯ is shaded.

◯ is not shaded.

Both parts together show _____ .

Adding Fractional Parts

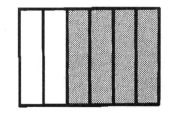

Shade $\frac{2}{5}$ of the circle.

$\left(\frac{3}{4}\right) + \left(\frac{1}{4}\right) = \frac{4}{4}$

⬭ + ⬭ =

⬭ + ⬭ =

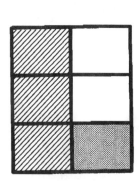

$\left(\frac{1}{6}\right) + \left(\frac{3}{6}\right) = \frac{4}{6}$

$\left(\frac{3}{6}\right) + \left(\frac{2}{6}\right) =$

$\left(\frac{1}{6}\right) + \left(\frac{3}{6}\right) + \left(\frac{2}{6}\right) =$

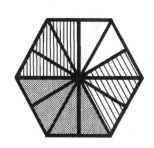

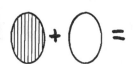

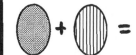

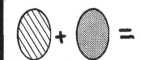

⬭ + ⬭ =

⬭ + ⬭ =

⬭ + ⬭ =

⬭ + ⬭ + ⬭ =

Try these problems without pictures.

$\frac{1}{3} + \frac{1}{3} =$ $\qquad$ $\frac{3}{7} + \frac{2}{7} =$ $\qquad$ $\frac{3}{8} + \frac{5}{8} =$ $\qquad$ $\frac{3}{5} + \frac{1}{5} =$

Fractions in Measurement

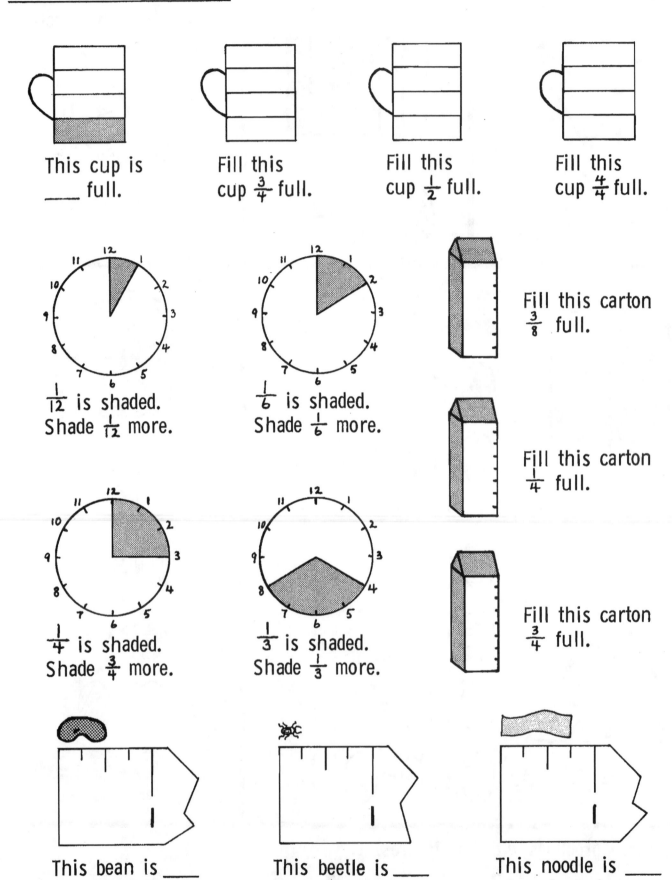

This cup is ____ full.

Fill this cup $\frac{3}{4}$ full.

Fill this cup $\frac{1}{2}$ full.

Fill this cup $\frac{4}{4}$ full.

$\frac{1}{12}$ is shaded. Shade $\frac{1}{12}$ more.

$\frac{1}{6}$ is shaded. Shade $\frac{1}{6}$ more.

Fill this carton $\frac{3}{8}$ full.

Fill this carton $\frac{1}{4}$ full.

$\frac{1}{4}$ is shaded. Shade $\frac{3}{4}$ more.

$\frac{1}{3}$ is shaded. Shade $\frac{1}{3}$ more.

Fill this carton $\frac{3}{4}$ full.

This bean is ____ of an inch long.

This beetle is ____ of an inch long.

This noodle is ____ of an inch long.

Fractions in Word Problems

<table>
<tr><td colspan="2" align="center">Sheila</td></tr>
<tr><td colspan="2" align="center">Math Quiz</td></tr>
<tr><td>1. d ✓</td><td>6. c ✗</td></tr>
<tr><td>2. b ✗</td><td>7. i ✓</td></tr>
<tr><td>3. a ✓</td><td>8. K ✓</td></tr>
<tr><td>4. f ✓</td><td>9. g ✓</td></tr>
<tr><td>5. h ✗</td><td>10. j ✓</td></tr>
</table>

There are __10__ answers on the paper.

__7__ of the __10__ answers are correct.

What fraction of the answers are correct? $\frac{7}{10}$

There are ____ books in the group.

____ of the ____ books are open.

What fraction of the books are open? ____

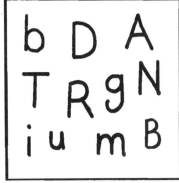

There are ____ letters in the group.

____ of the ____ letters are vowels.

What fraction of the letters are vowels? ____

____ of the ____ letters are capitals.

What fraction of the letters are capitals? ____

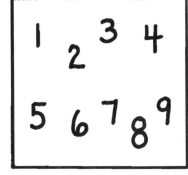

There are ____ numbers in the group.

____ of the ____ numbers are odd.

What fraction of the numbers are odd? ____

____ of the ____ numbers are even.

What fraction of the numbers are even? ____

14

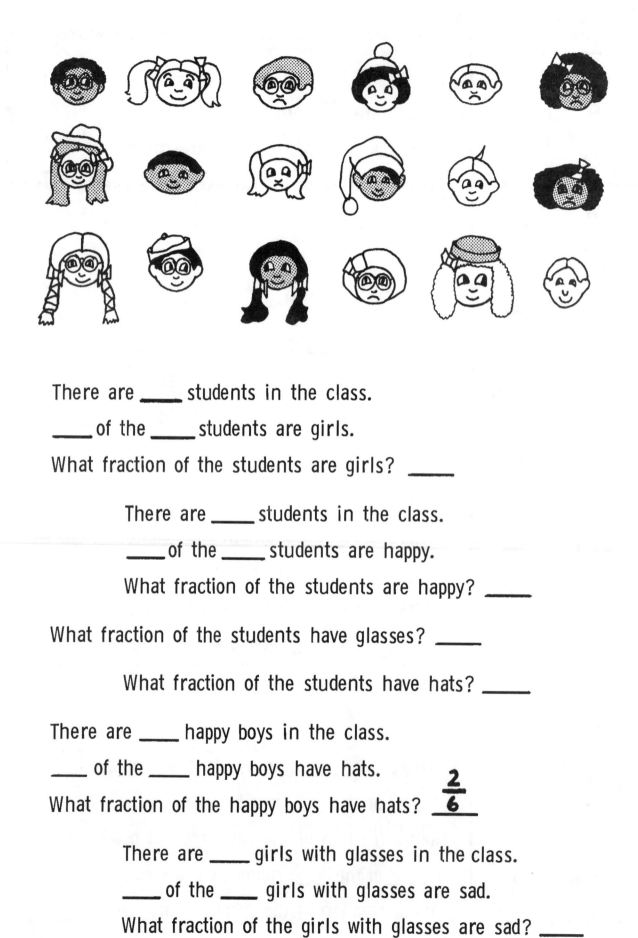

There are ____ students in the class.

____ of the ____ students are girls.

What fraction of the students are girls? ____

There are ____ students in the class.

____ of the ____ students are happy.

What fraction of the students are happy? ____

What fraction of the students have glasses? ____

What fraction of the students have hats? ____

There are ____ happy boys in the class.

____ of the ____ happy boys have hats.

What fraction of the happy boys have hats? $\frac{2}{6}$

There are ____ girls with glasses in the class.

____ of the ____ girls with glasses are sad.

What fraction of the girls with glasses are sad? ____

add	circle	count	cube	decimal	divide	divisor	graph
line	meter	number	plane	point	set	square	zero

There are _____ words in the list.

_____ of the _____ words begin with c.

What fraction of the words begin with c? _____

What fraction of the words begin with p? _____

What fraction of the words end with e? _____

What fraction of the words have a t? _____

What fraction of the words have exactly four letters? _____

There are _____ words that have exactly five letters.

_____ of the _____ five letter words begin with p?

What fraction of the five letter words begin with p? _____

There are _____ words that end with r.

_____ of the _____ words that end with r begin with d.

What fraction of the words that end with r begin with d? _____

What fraction of the six letter words begin with c? _____

What fraction of the words that begin with s have an e? _____

Fraction Vocabulary

The top and bottom numerals in a fraction have names. The top is called the <u>numerator</u> of the fraction and the bottom is called the <u>denominator</u> of the fraction. The little line that separates the numerator and denominator is called the <u>fraction bar</u>.

numerator → $\dfrac{3}{4}$ ← fraction bar

denominator →

Write the fraction.

8 is the numerator; 20 is the denominator. The fraction is $\dfrac{8}{20}$.

6 is the numerator; 7 is the denominator. The fraction is _____.

3 is the numerator; 4 is the denominator. The fraction is _____.

28 is the denominator; 10 is the numerator. The fraction is _____.

2 is the denominator; 1 is the numerator. The fraction is _____.

7 is the numerator; 8 is the denominator. The fraction is _____.

6 is the denominator; 0 is the numerator. The fraction is _____.

Fill in the blanks.

In $\frac{3}{8}$, 8 is the __denominator__ and 3 is the _____ .

In $\frac{5}{6}$, 5 is the _____ and 6 is the _____ .

In $\frac{1}{7}$, 1 is the _____ and 7 is the _____ .

In $\frac{20}{35}$, 35 is the _____ and 20 is the _____ .

In $\frac{1}{50}$, 1 is the _____ and 50 is the _____ .

Reading and Writing Fractions

Match.

$\overset{\frown}{20}\ \overset{\frown}{20}$	$\cancel{\dfrac{11}{15}}$	$\dfrac{18}{12}$	$\dfrac{3}{50}$	100
$\dfrac{7}{11}$	$\dfrac{12}{18}$	$11:15$	28	$\dfrac{20}{8}$
1980	$\$3.50$	$\dfrac{11}{10}$	ELEVEN	$\dfrac{20}{20}$
$\dfrac{1}{100}$	3×4	$\dfrac{3}{4}$	$\dfrac{19}{80}$	$\dfrac{10}{11}$

$\dfrac{11}{15}$ eleven fifteenths

_____ one hundred

_____ twenty twenty

_____ twenty eighths

_____ three fifty

_____ eighteen twelfths

_____ twelve eighteenths

_____ three fiftieths

_____ nineteen eighty

_____ ten elevenths

_____ twenty-eight

_____ nineteen eightieths

_____ seven elevenths

_____ seven eleven

_____ eleven fifteen

_____ twenty twentieths

_____ eleven tenths

_____ one hundredth

_____ three fours

_____ three fourths

This short article appeared in the school newspaper. Underline every fraction.

The girls basketball team won the league title this season. The team won seven eighths of their sixteen games. Next year might be a tough one though because three fourths of the twelve girls on the team are seniors. The boys basketball team had a fine season also, winning three fifths of their home games and half of their away games. Things look good for the boys next year because twelve of their fifteen players will be back.

Write the numeral for each fraction.

one fifth	$\frac{1}{5}$	eight twelfths	
one eighth		ten elevenths	
one twelfth		thirteen fourteenths	
two thirds		thirteen fortieths	
two sixths		thirteen forty-fourths	
three seventeenths		twenty twenty-sevenths	
four fourths		twenty-seven thirtieths	
four tenths		thirty-four fiftieths	
four elevenths		fifty hundredths	
five nineteenths		fifty-six sixtieths	
five thirty-eighths		eighty-nine ninetieths	
six twentieths		one hundred hundredths	

Write the numeral for the underlined words.

The class was <u>three fourths</u> of an hour long. _____

Phil spent <u>one half</u> of a dollar. _____

Ms. Harris spent <u>one fourth</u> of her income on rent. _____

Mr. Garcia read <u>two thirds</u> of the book. _____

Judy walked <u>six tenths</u> of a kilometer to school. _____

<u>Two fifths</u> of the windows were broken. _____

Write the fractions below just as you would say them aloud. The list at the side of the page will help you spell them correctly.

$\frac{1}{2}$ ___one half___ $\frac{3}{7}$ ___three sevenths___   one
two
three
four
five
$\frac{2}{3}$ _____ $\frac{3}{8}$ _____   six
seven
eight
$\frac{1}{4}$ _____ $\frac{3}{9}$ _____   nine
ten
$\frac{2}{4}$ _____ $\frac{4}{5}$ _____   eleven
twelve
twenty
thirty
$\frac{2}{5}$ _____ $\frac{3}{10}$ _____
half
third
$\frac{5}{6}$ _____ $\frac{6}{10}$ _____   fourth
fifth
$\frac{4}{7}$ _____ $\frac{3}{3}$ _____   sixth
seventh
eighth
$\frac{6}{8}$ _____ $\frac{6}{6}$ _____   ninth
tenth
eleventh
$\frac{5}{11}$ _____ $\frac{1}{23}$ _____   twelfth
twentieth
thirtieth
$\frac{1}{12}$ _____ $\frac{11}{12}$ _____
halves
thirds
$\frac{1}{20}$ _____ $\frac{8}{20}$ _____   fourths
fifths
sixths
$\frac{1}{30}$ _____ $\frac{10}{30}$ _____   sevenths
eighths

Fractions Equal to One

Fill in the tags.

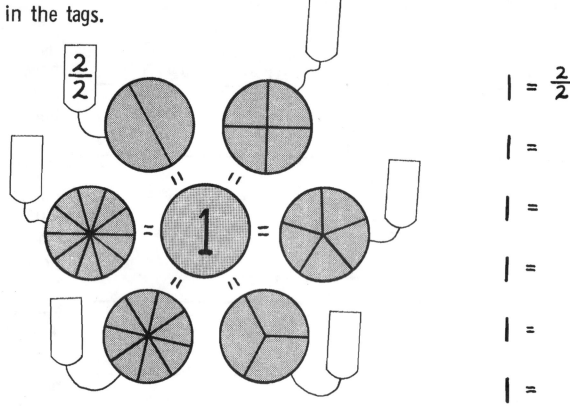

$|$ = $\frac{2}{2}$

$|$ =

$|$ =

$|$ =

$|$ =

$|$ =

We can say that: $|$ = $\frac{2}{2}$ = ___ = ___ = ___ = ___ .

Some other fraction names for one are: $\frac{6}{6}$, $\frac{7}{7}$, $\frac{11}{11}$, $\frac{15}{15}$, $\frac{23}{23}$.

List five more fraction names for one: ___ , ___ , ___ , ___ , ___ .

Circle the fraction equal to one.	Circle the fraction equal to $\frac{3}{3}$.	Circle the fraction equal to $\frac{10}{10}$.
$\frac{2}{3}$ $\frac{4}{4}$ $\frac{7}{8}$ $\frac{1}{2}$	$\frac{4}{5}$ $\frac{1}{7}$ $\frac{2}{5}$ $\frac{8}{8}$	$\frac{2}{7}$ $\frac{1}{8}$ $\frac{2}{2}$ $\frac{10}{11}$

Circle all of the fractions equal to one.

$\frac{11}{111}$ $\frac{15}{23}$ $\frac{5}{6}$ $\frac{18}{18}$ $\frac{3}{10}$ $\frac{1}{6}$ $\frac{100}{100}$ $\frac{7}{8}$ $\frac{1}{1}$

$\frac{6}{101}$ $\frac{6}{6}$ $\frac{1}{12}$ $\frac{3}{4}$ $\frac{2}{2}$ $\frac{5}{12}$

Fractions and Number Lines

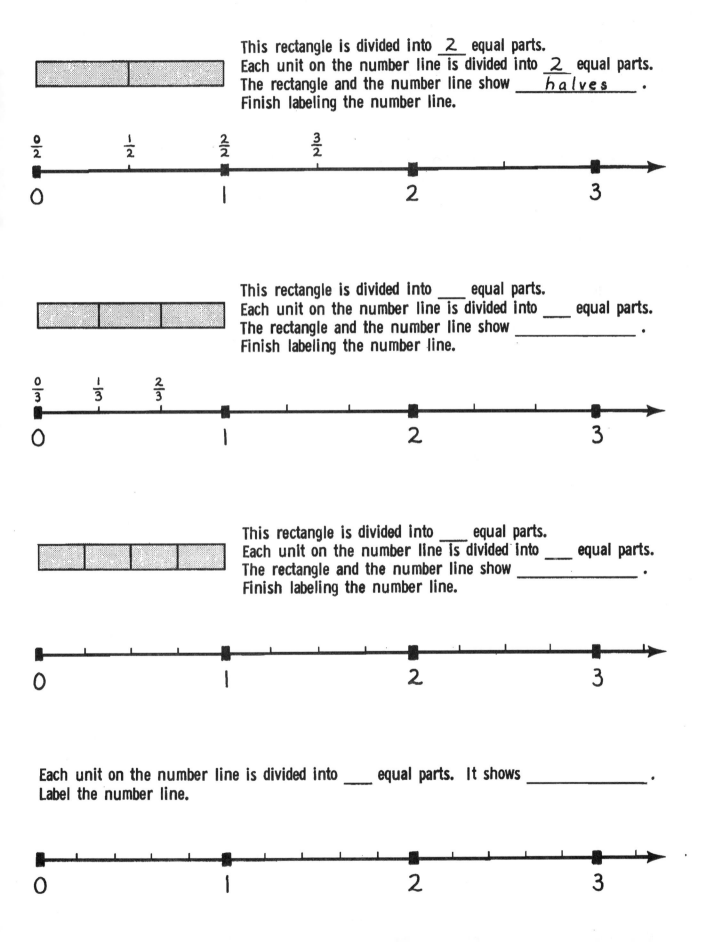

This rectangle is divided into _2_ equal parts.
Each unit on the number line is divided into _2_ equal parts.
The rectangle and the number line show _halves_.
Finish labeling the number line.

$\frac{0}{2}$ $\frac{1}{2}$ $\frac{2}{2}$ $\frac{3}{2}$

0   1   2   3

This rectangle is divided into ___ equal parts.
Each unit on the number line is divided into ___ equal parts.
The rectangle and the number line show _____.
Finish labeling the number line.

$\frac{0}{3}$ $\frac{1}{3}$ $\frac{2}{3}$

0   1   2   3

This rectangle is divided into ___ equal parts.
Each unit on the number line is divided into ___ equal parts.
The rectangle and the number line show _____.
Finish labeling the number line.

0   1   2   3

Each unit on the number line is divided into ___ equal parts. It shows _____.
Label the number line.

0   1   2   3

A number line can be labeled with fractions or with whole numbers and mixed numbers. (A <u>mixed number</u> is a whole number together with a fraction.) Finish labeling the number lines. Write a fraction above each mark and write a whole number or mixed number below each mark.

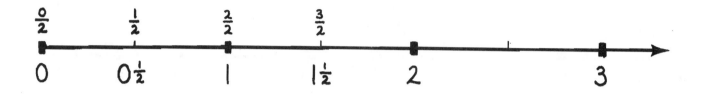

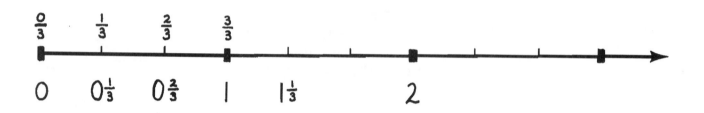

Write the whole or mixed number that equals each fraction. Use the number lines above.

$$\frac{3}{2} = 1\frac{1}{2} \qquad \frac{5}{2} = \qquad \frac{2}{2} = \qquad \frac{6}{2} =$$

$$\frac{4}{3} = \qquad \frac{5}{3} = \qquad \frac{8}{3} = \qquad \frac{9}{3} =$$

$$\frac{5}{4} = \qquad \frac{11}{4} = \qquad \frac{7}{4} = \qquad \frac{13}{4} =$$

Fractions Greater than One

Fraction

Each circle is divided into halves.
There are __9__ halves shaded.

$\dfrac{9}{2}$ of the circles are shaded.

Mixed Number

__4__ circles are completely shaded.
$\dfrac{1}{2}$ of another circle is shaded.

$4\dfrac{1}{2}$ circles are shaded.

Each hexagon is divided into sixths.
There are _____ sixths shaded.

_____ of the hexagons are shaded.

_____ hexagons are completely shaded.
_____ of another hexagon is shaded.

_____ hexagons are shaded.

Each square is divided into fourths.
There are _____ fourths shaded.

_____ of the squares are shaded.

_____ squares are completely shaded.
_____ of another square is shaded.

_____ squares are shaded.

$\dfrac{11}{3} =$

$\dfrac{19}{4} =$

Quarters

$\dfrac{15}{4} =$

Half Dollars

$= 3$

Comparing Fractions

Shade the **squares** and then put a loop around the correct answer.

Shade $\frac{1}{2}$.   Shade $\frac{1}{3}$.

   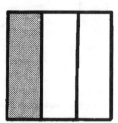

$\frac{1}{2}$ ⟨has more shading than⟩
    has as much shading as $\frac{1}{3}$
    has less shading than

Shade $\frac{1}{4}$.   Shade $\frac{1}{3}$.

$\frac{1}{4}$ has more shading than
    has as much shading as $\frac{1}{3}$
    has less shading than

Shade $\frac{2}{4}$.   Shade $\frac{1}{2}$.

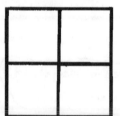   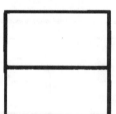

$\frac{2}{4}$ has more shading than
    has as much shading as $\frac{1}{2}$
    has less shading than

Shade $\frac{2}{3}$.   Shade $\frac{3}{4}$.

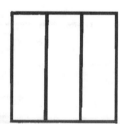   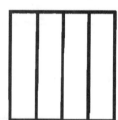

$\frac{2}{3}$ has more shading than
    has as much shading as $\frac{3}{4}$
    has less shading than

Shade $\frac{2}{3}$.   Shade $\frac{1}{2}$.

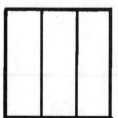   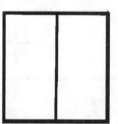

$\frac{2}{3}$ is greater than
    is equal to $\frac{1}{2}$
    is less than

Shade $\frac{4}{5}$.   Shade $\frac{3}{5}$.

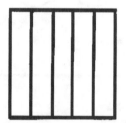   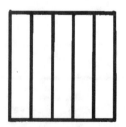

$\frac{4}{5}$ is greater than
    is equal to $\frac{3}{5}$
    is less than

Shade the squares.   Then fill in the blank in one of the following ways:

is greater than
is equal to
is less than

Shade $\frac{1}{6}$.

$\frac{1}{6}$ _____

Shade $\frac{1}{4}$.

$\frac{1}{4}$

Shade $\frac{2}{3}$.

$\frac{2}{3}$ _____

Shade $\frac{2}{10}$.

$\frac{2}{10}$

Shade $\frac{3}{5}$.

$\frac{3}{5}$ _____

Shade $\frac{3}{10}$.

$\frac{3}{10}$

Shade $\frac{3}{6}$.

$\frac{3}{6}$ _____

Shade $\frac{2}{4}$.

$\frac{2}{4}$

Shade $\frac{3}{4}$.

$\frac{3}{4}$ _____

Shade $\frac{1}{4}$.

$\frac{1}{4}$

Shade $\frac{1}{2}$.

$\frac{1}{2}$ _____

Shade $\frac{3}{4}$.

$\frac{3}{4}$

Shade the squares. Then use **>** , **=** , or **<** to make each statement true.

> **>** means "is greater than"
>
> **=** means "is equal to"
>
> **<** means "is less than"

Shade $\frac{2}{3}$.

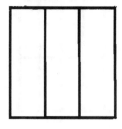

$\frac{2}{3}$

Shade $\frac{4}{6}$.

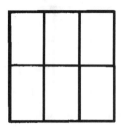

$\frac{4}{6}$

Shade $\frac{1}{5}$.

$\frac{1}{5}$

Shade $\frac{1}{2}$.

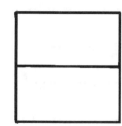

$\frac{1}{2}$

Shade $\frac{3}{4}$.

$\frac{3}{4}$

Shade $\frac{3}{9}$.

$\frac{3}{9}$

Shade $\frac{6}{9}$.

$\frac{6}{9}$

Shade $\frac{2}{3}$.

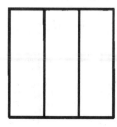

$\frac{2}{3}$

Shade $\frac{2}{5}$.

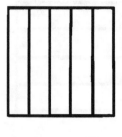

$\frac{2}{5}$

Shade $\frac{5}{8}$.

$\frac{5}{8}$

Shade $\frac{3}{6}$.

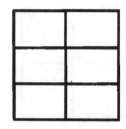

$\frac{3}{6}$

Shade $\frac{1}{4}$.

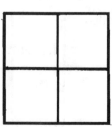

$\frac{1}{4}$

Write the fraction for the shaded part of each rectangle.

$\frac{3}{8}$ ___ ___ ___ ___ ___ ___ ___ ___

Rearrange the fractions above from smallest to largest.

___ ___ $\frac{2}{8}$ ___ ___ ___ ___ ___ ___
smallest   largest

For fractions with the same denominator:

As the numerators get larger, the fractions get _____.
(larger/smaller)

As the numerators get smaller, the fractions get _____.
(larger/smaller)

The smallest fraction is the fraction with the smallest _____.
(numerator/denominator)

The _____ fraction is the fraction with the largest _____.
(largest/smallest)   (numerator/denominator)

Write the fraction for the shaded part of each circle.

___ ___ ___ ___ ___ ___ ___ ___

Rearrange the fractions above from smallest to largest.

___ ___ ___ ___ ___ ___ ___ ___
smallest   largest

For fractions with the same numerator:

As the denominators get smaller, the fractions get _____.
(larger/smaller)

As the denominators get larger, the fractions get _____.

The smallest fraction is the fraction with the _____ denominator.

The largest fraction is the fraction with the _____ _____.

Comparing Fractions Using Number Lines

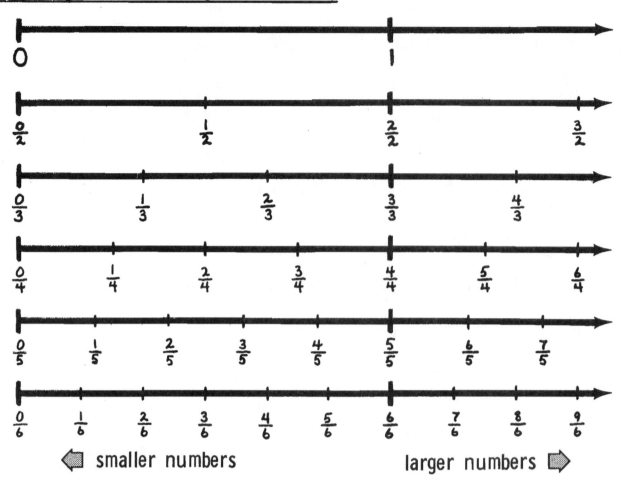

⬅ smaller numbers   larger numbers ⮕

To do each problem below:

1. Find both fractions on the number lines.
2. Put a finger on each.
3. Decide which fraction is larger and which is smaller or if both are equal.
4. Put **>** , **<** , or **=** between the fractions to make a true statement.

$\frac{2}{5}$	**>**	$\frac{1}{4}$	$\frac{2}{5}$	$\frac{3}{4}$	$\frac{2}{5}$ 1
$\frac{4}{6}$		$\frac{2}{3}$	$\frac{4}{6}$	$\frac{3}{2}$	$\frac{4}{4}$ $\frac{2}{3}$
$\frac{6}{5}$		$\frac{5}{6}$	$\frac{3}{2}$	$\frac{2}{3}$	$\frac{8}{6}$ $\frac{7}{5}$
$\frac{0}{2}$		0	$\frac{0}{4}$	0	0 $\frac{0}{6}$

Equal Fractions

Write the fraction for the shaded part of each figure below.

Equal parts shaded.	Equal parts shaded.
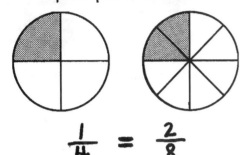 $\frac{1}{4} = \frac{2}{8}$	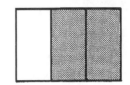 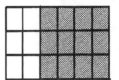 =
Equal parts shaded.	Equal parts shaded.
  =	=
Shade equal parts.	Shade equal parts.
 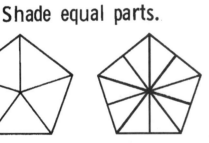 =	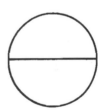  =
Shade equal parts.	Shade equal parts.
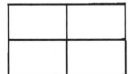  =	 =

$\frac{1}{2}$ shaded.

Fractions equal to $\frac{1}{2}$.

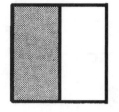
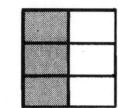
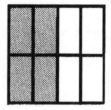
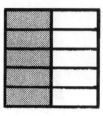

$\frac{1}{2}$

———   ———   ———   ———

We can say that: $\frac{1}{2}$ =   =   =   =   .

$\frac{3}{4}$ shaded.

Shade fractions equal to $\frac{3}{4}$.

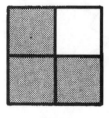

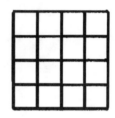
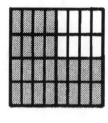

$\frac{3}{4}$

———   ———   ———   ———

We can say that: $\frac{3}{4}$ =   =   =   =   .

Shade $\frac{1}{3}$.

Shade fractions equal to $\frac{1}{3}$.

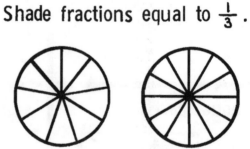

$\frac{1}{3}$

———   ———   ———   ———

We can say that: $\frac{1}{3}$ =   =   =   =

Finding Equal Fractions

Below is another way to find equal fractions. Start with any fraction. Pick a number larger than one. Multiply the numerator and denominator of the fraction by that number. The fraction you make looks different than the fraction you started with, but it has the same value. You have found an equal fraction.

Pick 2:
$$\frac{1^{\times 2}}{3_{\times 2}} = \frac{2}{6}$$

Pick 3:
$$\frac{1^{\times 3}}{3_{\times 3}} = \frac{3}{9}$$

Pick 4:
$$\frac{1^{\times 4}}{3_{\times 4}} = \frac{4}{12}$$

Now you find some fractions equal to $\frac{1}{3}$.

Pick 5:
$$\frac{1^{\times 5}}{3_{\times 5}} =$$

Pick 6:
$$\frac{1^{\times 6}}{3_{\times 6}} =$$

Pick 7:
$$\frac{1^{\times 7}}{3_{\times 7}} =$$

Pick 10:
$$\frac{1^{\times 10}}{3_{\times 10}} =$$

Pick 20:
$$\frac{1^{\times 20}}{3_{\times 20}} =$$

Find some fractions equal to $\frac{2}{3}$.

Pick 2:
$$\frac{2^{\times 2}}{3_{\times 2}} =$$

Pick 3:
$$\frac{2^{\times 3}}{3_{\times 3}} =$$

Pick 4:
$$\frac{2^{\times 4}}{3_{\times 4}} =$$

Pick 5:
$$\frac{2}{3} =$$

Pick 6:
$$\frac{2}{3} =$$

Find some fractions equal to $\frac{1}{4}$.

Pick 2:
$$\frac{1}{4} =$$

Pick 3:
$$\frac{1}{4} =$$

Pick 4:
$$\frac{1}{4} =$$

Pick ☐:
$$\frac{1}{4} =$$

Pick ☐:
$$\frac{1}{4} =$$

Think about it ...

$$\frac{1}{3} = \frac{2}{6} \quad \text{because} \quad \frac{1}{3} = \frac{1}{3} \times \boxed{1} = \frac{1}{3} \times \boxed{\frac{2}{2}} = \frac{1^{\times 2}}{3_{\times 2}} = \frac{2}{6}$$

$$\frac{1}{4} = \frac{5}{20} \quad \text{because} \quad \frac{1}{4} = \frac{1}{4} \times \boxed{1} = \frac{1}{4} \times \boxed{\frac{5}{5}} = \frac{1^{\times 5}}{4_{\times 5}} = \frac{5}{20}$$

On this page you must make strings of equal fractions. To make a string of equal fractions you pick a number (larger than 1), multiply, and make an equal fraction. Then you pick another number, multiply, and make another equal fraction. Keep picking numbers and multiplying (always by the numerator and denominator of the first fraction) until you have finished the string.

	Pick 2:	Pick 3:	Pick 4:	Pick 5:	Pick 6:

$$\frac{3}{4} = \frac{6}{8} = \frac{9}{12} = \frac{12}{16} = \frac{15}{20} = \frac{18}{24}$$

Find five fractions equal to $\frac{1}{2}$.

	Pick 2:	Pick 3:	Pick 4:	Pick 5:	Pick □:

$$\frac{1}{2} = \quad = \quad = \quad = \quad$$

Find five fractions equal to $\frac{1}{5}$. You pick all the numbers to multiply by.

$$\frac{1}{5} = \quad = \quad = \quad = \quad$$

Find five fractions equal to $\frac{2}{5}$.

$$\frac{2}{5} = \quad = \quad = \quad = \quad$$

33

Find four fractions equal to $\frac{3}{5}$.

Pick 2:   Pick 3:   Pick 4:   Pick 5:

$$\frac{3}{5} = \frac{6}{10} = \frac{9}{15} = \frac{12}{20} = \frac{15}{25}$$

Find four fractions equal to $\frac{1}{4}$.

$$\frac{1}{4} = \quad = \quad = \quad =$$

Find four fractions equal to $\frac{1}{8}$.

$$\frac{1}{8} = \quad = \quad = \quad =$$

Find four fractions equal to $\frac{3}{8}$.

$$\frac{3}{8} = \quad = \quad = \quad =$$

Find four fractions equal to $\frac{5}{8}$.

$$\frac{5}{8} = \quad = \quad = \quad =$$

Find four fractions equal to $\frac{4}{9}$.

$$\frac{4}{9} = \quad = \quad = \quad =$$

Find four fractions equal to $\frac{1}{10}$.

$$\frac{1}{10} = \quad = \quad = \quad =$$

Find four fractions equal to $\frac{3}{10}$.

$$\frac{3}{10} = \quad = \quad = \quad =$$

Remember, when you multiply the numerator and the denominator of a fraction by the same number (larger than 1) you make an equal fraction.

Pick **5**:

$$\frac{2 \times 5}{3 \times 5} =$$

Pick ☐:

$$\frac{4}{5} =$$

Pick ☐:

$$\frac{1}{8} =$$

Make equal fractions. First figure out what the numerator of the fraction was multiplied by and then multiply the denominator by the same number.

$3 \times \boxed{5} = 15$ so you must pick $\boxed{5}$.

$$\frac{3 \times 5}{4 \times 5} = \frac{15}{20}$$

$$\frac{5}{8} = \frac{15}{}$$

$$\frac{1}{3} = \frac{5}{}$$

$$\frac{2}{5} = \frac{4}{}$$

$$\frac{3}{20} = \frac{9}{}$$

$$\frac{1}{2} = \frac{10}{}$$

$$\frac{3}{4} = \frac{18}{}$$

$5 \times \square = 10$ so you must pick ☐.

$$\frac{5}{8} = \frac{10}{}$$

$$\frac{1}{7} = \frac{4}{}$$

$$\frac{3}{7} = \frac{18}{}$$

$$\frac{7}{50} = \frac{14}{}$$

$$\frac{3}{5} = \frac{21}{}$$

$$\frac{2}{9} = \frac{10}{}$$

$$\frac{5}{8} = \frac{50}{}$$

$1 \times \square = 3$ so you must pick ☐.

$$\frac{1}{7} = \frac{3}{}$$

$$\frac{3}{7} = \frac{6}{}$$

$$\frac{9}{10} = \frac{90}{}$$

$$\frac{11}{25} = \frac{44}{}$$

$$\frac{8}{8} = \frac{80}{}$$

$$\frac{5}{12} = \frac{20}{}$$

$$\frac{5}{7} = \frac{40}{}$$

Find the missing numerators to make equal fractions.

$$7 \times \boxed{2} = 14 \quad so$$
you must pick $\boxed{2}$.

$$\frac{6^{\times 2}}{7^{\times 2}} = \frac{}{14} \qquad \frac{4}{9} = \frac{}{36} \qquad \frac{7}{13} = \frac{}{26}$$

$$\frac{1}{5} = \frac{}{10} \qquad \frac{1}{25} = \frac{}{75} \qquad \frac{3}{11} = \frac{}{66}$$

$$\frac{5}{5} = \frac{}{40} \qquad \frac{1}{20} = \frac{}{40} \qquad \frac{5}{6} = \frac{}{30}$$

$$\frac{4}{10} = \frac{}{30} \qquad \frac{2}{11} = \frac{}{55} \qquad \frac{2}{9} = \frac{}{63}$$

Find the missing numerators or denominators.

$$\frac{2}{5} = \frac{10}{} \qquad \frac{2}{7} = \frac{}{42} \qquad \frac{4}{16} = \frac{}{32}$$

$$\frac{5}{6} = \frac{}{18} \qquad \frac{4}{9} = \frac{12}{} \qquad \frac{2}{7} = \frac{14}{}$$

$$\frac{2}{9} = \frac{}{45} \qquad \frac{2}{11} = \frac{}{33} \qquad \frac{5}{9} = \frac{20}{}$$

$$\frac{1}{6} = \frac{7}{} \qquad \frac{5}{12} = \frac{25}{} \qquad \frac{15}{15} = \frac{}{45}$$

$$\frac{3}{8} = \frac{24}{} \qquad \frac{1}{8} = \frac{5}{} \qquad \frac{0}{3} = \frac{}{9}$$

Practice Test - <u>Key To Fractions</u> Book 1

Name _____

Date _____

Which shows fourths?

The rectangle is divided into two equal parts or

_____ .

The square is divided into three parts.

It $\frac{\text{does}}{\text{does not}}$ show thirds.

Use a fraction to name the shaded part of each figure.

Shade $\frac{3}{5}$.

Fill this carton $\frac{3}{4}$ full.

Finish the problem below.

$\frac{2}{6}$ + ◯ + ◯ =

---

7 is the denominator; 3 is the numerator.   The fraction is ____ .

In $\frac{5}{12}$, ____ is the numerator and ____ is the denominator.

There are ____ figures in the group.

____ of the ____ figures are shaded.

What fraction of the figures are shaded? ____

What fraction of the shaded figures are squares? ____

What fraction of the triangles are shaded? ____

Practice Test - Page 2

Write the numeral.

one tenth _____   five eighths _____

Circle the fractions equal to one.

$\frac{1}{2}$ $\frac{5}{6}$ $\frac{4}{40}$ $\frac{8}{8}$ $\frac{11}{16}$ $\frac{9}{11}$

$\frac{2}{2}$ $\frac{7}{15}$ $\frac{1}{6}$ $\frac{5}{7}$ $\frac{100}{100}$ $\frac{1}{1}$

Finish labeling the number line. Write a fraction above each mark and write a whole number or a mixed number below each mark.

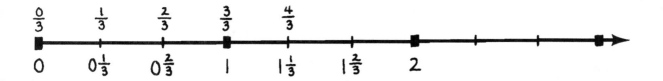

$\frac{0}{3}$ $\quad$ $\frac{1}{3}$ $\quad$ $\frac{2}{3}$ $\quad$ $\frac{3}{3}$ $\quad$ $\frac{4}{3}$

0 $\quad$ $0\frac{1}{3}$ $\quad$ $0\frac{2}{3}$ $\quad$ 1 $\quad$ $1\frac{1}{3}$ $\quad$ $1\frac{2}{3}$ $\quad$ 2

Shade $\frac{1}{4}$.   Shade $\frac{5}{8}$.

$\frac{1}{4}$ is greater than
is equal to
is less than $\frac{5}{8}$

Shade equal parts.
Then write each fraction.

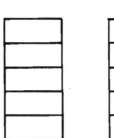

_____ = _____

Find equal fractions.

$\frac{1^{\times 2}}{2_{\times 2}} =$

$\frac{3^{\times 5}}{4_{\times 5}} =$

You pick a number to multiply by.

$\frac{2}{5} =$

Find four fractions equal to $\frac{1}{3}$.

Pick 2:   Pick 3:   Pick ☐:   Pick ☐:

$\frac{1}{3} =$ $\quad$ $=$ $\quad$ $=$ $\quad$ $=$

Make equal fractions.

$\frac{1}{2} = \frac{}{6}$ $\qquad$ $\frac{3}{5} = \frac{}{10}$ $\qquad$ $\frac{3}{4} = \frac{}{16}$

$\frac{3}{7} = \frac{9}{}$ $\qquad$ $\frac{3}{7} = \frac{12}{}$ $\qquad$ $\frac{5}{6} = \frac{10}{}$

Key to Fractions® workbooks

Book 1:   Fraction Concepts
Book 2:   Multiplying and Dividing
Book 3:   Adding and Subtracting
Book 4:   Mixed Numbers

Answers and Notes for Books 1–4
Reproducible Tests for Books 1–4

Also available in the Key to…® series

Key to Decimals®
Key to Percents®
Key to Algebra®
Key to Geometry®
Key to Measurement®
Key to Metric Measurement®
The Key to Tracker®, the online companion for the
Key to Fractions, Decimals, Percents, and Algebra workbooks

Chartwell-Yorke Ltd
114 High Street, Belmont Village,
Bolton, Lancashire, BL7 8AL, England
Tel: (+44) (0)1204 811001
Fax: (+44) (0)1204 811008
info@chartwellyorke.com
http://www.chartwellyorke.com

...iculum Press
...MATHEMATICS EDUCATION

ISBN 978-0-913684-91-7